GROWING UP WILD

How Young Animals Survive

BY RUSSELL FREEDMAN

drawings by Leslie Morrill

Holiday House

New York

FOR CLARK RUSSELL KORMIER

591.3
FRE

Library of Congress Cataloging in Publication Data

Freedman, Russell.
Growing up wild.

Bibliography: p. 64
SUMMARY: Describes the growth and development of six
animals from their birth until they become independent.
1. Animals, Infancy of — Juvenile literature.
2. Parental behavior in animals — Juvenile literature.
[1. Animals — Infancy. 2. Animals — Habits and behavior]
I. Morrill, Leslie H. II. Title.
QL763.F68   596'.03'9   75-10795
ISBN 0-8234-0265-7

Contents

During the first few months of its life, a baby gorilla never leaves its mother's side. It rides on her back until it is two or three years old. It sleeps with her or near her until it is four or older. As the youngster grows up it is protected and pampered not only by its mother, but by all the gorillas in its family.

A tadpole never sees its parents. When it wriggles out of its jelly-covered egg, it is weak, defenseless, and likely to be eaten. Its mother deserted her eggs soon after laying them. Frogs do not usually care for their young. They flourish and multiply because each female lays hundreds or thousands of eggs.

Like the tadpole, most fishes, amphibians, and reptiles start life entirely on their own. They grow up without help from anyone. To survive they must depend on their instincts and luck. Like the gorilla, on the other hand, mammals and birds need the help and protection of one or both parents. As a rule, these animals cannot survive on their own.

This book shows how four young animals with parents and two young animals without parents meet the challenges of growing up in the wild.

From Tadpole to Leopard Frog

She could be seen moving inside her jelly-covered egg — a small dark speck of life, not much bigger than a comma.

At first she simply twitched now and then. As she grew larger and stronger she felt more and more cramped, and she squirmed almost constantly. Finally she broke through the jelly that surrounded her. She escaped from her egg and swam into the pond — head up, tail down.

She appeared to be all head and tail. Nothing about her suggested that she might become a leopard frog. In fact, she was barely recognizable as a tadpole. Her newly hatched body was not yet fully formed. She was quite flat and about one-fourth of an inch long. She had no eyes, no nostrils and no mouth.

As she wriggled across the April pond, she bumped

against a water weed. In front of her head, where her mouth would develop, she had two tiny suction cups. They released a sticky substance, and she fastened herself to the weed.

Without a mouth she couldn't feed. Part of the yolk from her egg was stored in her belly, and that nourished her. She breathed by means of feathery gills on either side of her head. For a few days she spent most of her time clinging to plants and resting. Sometimes she swam about aimlessly. She wasn't able to do much else.

Around her the water teemed with newly hatched tadpoles. Frogs had left masses of jelly-covered eggs all over the pond, abandoning them to hatch by themselves. Her own mother had laid three thousand bead-like eggs. Many had failed to develop. Many more were eaten by water insects and other creatures. Yet large numbers of eggs survived.

Hundreds of sisters and brothers had hatched with her. Thousands of other tadpoles were hatching every day. They swarmed into the pond, attaching themselves to weeds and rocks and globs of egg jelly.

Without fully developed eyes, she couldn't see the others all around her. She couldn't smell or hear them. But she could feel the water moving nearby.

Along her back and sides she had rows of tiny pits. Each pit contained a minute bristle which picked up the slightest vibrations in the water. At first these pits were her only link with the world outside her body. They helped her keep her balance as she swam, and they alerted her to danger. When she felt a sudden ripple in the water, she darted away.

She was constantly in danger of being swallowed by creatures larger than herself. Fish patrolled the pond in hungry packs. Snakes glided watchfully through the water and struck without warning. Snapping turtles lay hidden in the mud. And meat-eating insects lurked everywhere, waiting to ambush defenseless tadpoles.

Many tadpoles lost their lives soon after they hatched. But she escaped from her enemies and grew.

In a few days her body changed. Two small eyes developed. A pair of nostrils appeared. Her mouth opened, and her tiny suction cups disappeared.

A fold of skin grew over her outer gills, and they disappeared too. They were replaced by four pairs of new gills inside her body. She began to breathe as a fish does, by taking water into her mouth. As the water passed through her gill chamber, the gills absorbed oxygen. Then the water flowed out through a hole, the spiracle, on the left side of her body.

By the time she was a week old, she really looked like a tadpole. Her brownish body had nearly doubled in size. She was getting plump, and her broad fish-like tail was growing longer. Above and below her tail were shimmering black-speckled crests, similar to fins. The skin on her cream-colored belly was so thin that her tightly coiled intestine could be seen inside.

Now she spent most of her time eating. Her small round mouth was hard and tough. It was surrounded by rows of tiny toothlike ridges. She nibbled at water plants and scraped algae from rocks and sticks.

With her newly formed eyes she saw light and shadows and moving shapes. This gave her a better chance to survive. She couldn't tell one enemy from another, but she recognized the danger signals. When a shadow fell over her, or when a large shape moved suddenly before her, she swam for her life.

One morning, as she fed with other tadpoles near the edge of the pond, she was almost captured by a wolf spider. The spider sat motionless on shore, watching the tadpoles eat. Suddenly it dashed out across the water on its eight hairy legs. It dove under, stabbed a tadpole with its poison fangs, dragged the victim back to shore and devoured it.

She fled with the others into deeper water. A few minutes later a diving beetle swooped down on her. She twisted and darted away, skimming over the bottom of the pond.

Every day she narrowly escaped from death. For two months she wriggled about the pond, growing fatter and longer. Early that summer, when she was about 3 inches long, her body began to change again. This time she underwent a complete change — a metamorphosis.

Two bumps appeared on either side of her body, near the base of her tail. The bumps grew larger. Soon two tiny hind legs broke through her skin. She dragged them behind her as she swam.

Gradually her legs became longer. As toes formed, they began to look like frogs' legs. Now she used them to help push herself through the water.

Two arms had been forming inside her body, but they did not appear until her legs were quite long. First her left arm emerged through her spiracle, or water hole. For a short time she swam about with one arm and two legs. Then her right arm broke through the skin on the other side of her body.

Once her arms had appeared, her body changed more

rapidly. Her arms and legs thickened. Her eyes grew larger and began to bulge. Her small round mouth spread across her face, developing teeth, strong jaws and a long tongue. Her skin became thicker and tougher. And her tail began to vanish.

Day after day her tail grew shorter. Because of changes taking place in her mouth and digestive system, she could no longer feed on plants and algae. Instead, she obtained nourishment by absorbing her own tail. Each day her changing body used part of her tail as food.

Meanwhile her fish-like gills were shrinking. Lungs were forming inside her gill chamber. She began to rise to the surface to gulp air. Finally she used her arms and legs to climb onto a dry rock. She sat there briefly before plunging back into the water.

After that she spent more and more time out of the water. Soon her tail was just a stump. She was ready to feed again, but she was no longer interested in plants.

One morning she spotted a water flea moving nearby. She swam closer to smell the flea. Then she lunged at it. Her mouth snapped open as she caught the flea with her tongue. From then on she snapped at any moving creature slow enough to catch and small enough to swallow.

By the middle of summer her arms and legs were fully developed. Her tail was gone. Lungs had taken the place of her gills. She had changed from a fish-like water creature to an air-breathing land animal, from a plant eater to a hunter, from a dark wriggling tadpole to a spotted leopard froglet. She was ready to leave the pond and start hopping about on land.

Without her tail she was very small, less than 2 inches long, but she was unmistakably a frog. She had a long slender body with a narrow waist and a pointed snout. Her dark brown spots were outlined in white against a green background. Her underparts were glistening white. Her mouth was enormous. And her bulging bronze eyes sparkled and glittered.

Though she looked like a miniature adult, she was still a youngster. She would not be ready to mate for another year or two. And she would grow to be twice her present size or more.

Most of her sisters and brothers had perished in the pond. Perhaps one tadpole out of twenty had survived to climb up on land. Even so, hundreds of froglets had emerged from the pond with her. Because they were so tiny, they were difficult to see. Their spotted markings blended with the light and shadows of the surrounding grass. They became visible only when they made a startled leap into the water.

At first she stayed close to the edge of the pond along with the others, feeding on insects and spiders. Usually she sat quietly and waited until a meal came within striking distance. If an insect sat still beneath her nose, she would ignore it. Unless it moved, she didn't recognize it as food. But if it did move she would lunge forward and catch it by flicking out her long sticky tongue.

She swallowed her victims alive and whole. Her teeth were too small to chew with, so she used them only to grip her prey. When she had trouble swallowing a big mouthful, she lowered her eyes into the roof of her mouth and forced the victim back toward her throat. Often she used both hands to help cram a struggling mouthful down.

She would strike instantly at any small moving object no matter what it was. Sometimes she made a mistake and snapped at windblown feathers or specks of dirt. When that happened, she spit the object out and tried again.

She relied on her eyes. They spotted food, and they alerted her to danger. Her ears helped warn her too. She had no outer ears, but she did have two middle ears, or eardrums, which picked up airborne sounds. Her eardrums were round membranes located behind her eyes, just beneath the surface of her skin. Any sudden noise would put her on guard. But she needed her eyes to interpret the sound.

When she saw a falling shadow or a large moving object, she automatically leaped away. Her powerful hind legs were her chief means of defense. Her hop was long and low. With a single jump she could leap ten times her own body length. Each jump carried her in a slightly different direction as she zigzagged her way to safety.

One day she failed to jump quickly enough. As she sat at the edge of the pond, a shadow fell over her. She leaped toward the water but landed in the middle of a net.

The net swept her away from the pond and carried her up into the air. Struggling frantically, she managed to free herself and jump to the ground. She landed at the feet of a boy who was collecting specimens for his science class.

He loomed over her, blocking her path to the water. Unable to escape, she inflated her body with air. She bowed her head, spread her arms and legs and sucked air into her lungs, making her body appear much larger. The boy hesitated. Then he grabbed for her.

As his hand seized her, she emptied her bladder, releasing

a flood of water on the ground. At the same time she screamed in pain. To human ears it sounded like a baby's cry. Startled, the boy dropped her and stepped back.

She leaped past him, but he grabbed her again. She squirmed and twisted violently as he tried to stuff her into a jar. Her skin is what finally saved her. Like all frogs she had special glands which kept her skin moist and lubricated. She was as slippery as a greased pig and just as difficult to hold. She slipped out of the boy's grasp and with one tremendous leap reached the safety of the water.

Other frogs around the pond were not always as lucky. From time to time she heard the short shrill scream of a froglet as it was ambushed by a snake, snapped up by a turtle or captured by some other enemy. Ducks preyed on the froglets in water. Crows, hawks and other birds hunted them on land. So did skunks, raccoons, rats, cats and dogs.

Because she was a leopard frog — also called a "meadow frog" — she did not remain by the pond for long. Soon she felt an urge to travel inland. With many other tiny leopard froglets, she hopped and jumped to a nearby meadow.

For the rest of the summer she lived in the meadow, keeping mostly to herself. She was active mainly at night and early in the morning, when she searched the dew-drenched grass for insects.

She avoided hot open spaces. The summer sun could kill her quickly by drying her skin and heating her body. She rarely strayed from tall shady grass and from the dampest parts of the meadow. On hot sunny days she submerged herself in puddles or found refuge in dark cool hiding places.

Like most animals, she was "cold-blooded." She could

not maintain a constant body temperature, as warm-blooded birds and mammals do. Instead, her temperature changed from hour to hour — rising and falling with the thermometer. The inside of her body became as warm or cold as the air and ground around her. For this reason she had to choose her surroundings carefully. Some frogs in the area died of heat exhaustion as they tried to cross plowed fields and open roads.

Her skin gave her some protection by changing color with the weather. Cool temperatures and dark skies caused her skin to darken. This kept her warmer, because her body absorbed more heat. On warm bright days her skin turned lighter, and this kept her cooler.

Besides changing color, she changed her skin regularly. As a young frog she outgrew her old skin every few days and shed it in one piece. When she was ready to molt she would open and shut her mouth, stretch and twist her body and hump her back. Her tissue-thin outer skin would crack and split down her back and belly. Then she would wriggle out of her old skin, sucking it into her mouth and swallowing it as she went along. Shedding took just a few minutes. And her new skin was bright and clean.

At the end of September she left the meadow and found her way back to her native pond. On warm autumn days she could be seen swimming in the water or basking on shore. But as the weather grew colder, she spent more and more time near the bottom of the pond. Cold weather made her feel sluggish. The colder it got, the more slowly she moved.

Finally she burrowed into the mud beneath the water. She dug deeply, sinking down until mud had closed in all around her.

She could no longer breathe with her lungs. Instead she breathed with her skin, which had many tiny blood vessels close to the surface. Oxygen from the surrounding mud seeped through her skin and entered her bloodstream. She could obtain only a small amount of oxygen this way. But it was enough to keep her alive as she drifted into the deep sleep of hibernation.

Soon she stopped moving. Her heart slowed until it was barely beating. Her body turned cold. She seemed dead, but she wasn't. For the first time in her life, she was safe. Enemies could not reach her here, and she had burrowed deeply enough to keep from freezing.

Only a few of her sisters and brothers had survived. They slept near her beneath the chill muck of the winter pond, waiting for the warming temperatures of spring.

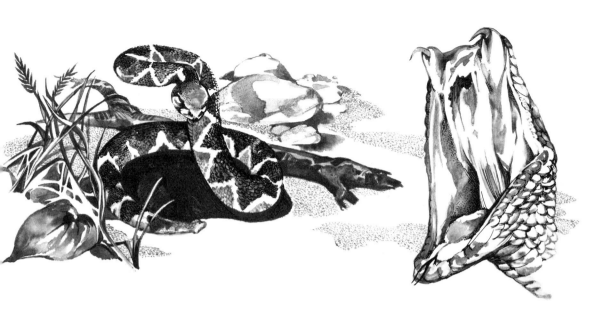

Timber Rattlesnake

He slipped out of his mother's body and lay coiled on the ground. He was wrapped in a transparent sac — a thin membrane that glistened and quivered.

Twisting and squirming, he broke the sac. He lifted his head, flicked out his tongue and gazed around. Then he crawled across his mother's back and stretched out to bask in the early morning sun.

Eleven others were born with him on that rocky mountain ridge. They wriggled and writhed all around their mother, but she paid no attention to them. She gave no sign that she even recognized them as her own. Finally she crawled away. Like all mother snakes, she abandoned her offspring and left them on their own.

The August sun grew hot and uncomfortable. Seeking

shade, he slithered under an overhanging rock. When he finally came out, his brothers and sisters were gone. He didn't miss them because he didn't remember them.

Late that afternoon he left the mountain ridge behind him. Slowly he made his way down the mountainside, weaving in and out among the rocks. His broad flat belly scales gripped the ground like tractor treads. Muscles rippled across his body as he pushed himself forward. Without arms or legs he could not travel very fast.

Every so often he stopped to test his surroundings. Lifting his wedge-shaped head, he flicked his forked tongue out through a notch in his upper lip. Then he went on his way again.

He was 9 inches long and as slim as a man's finger. Diamond-shaped scales covered his back and sides, forming dark brown crossbands against a yellow background. The scales on his belly were wider, and each scale overlapped the one behind. At the tip of his tail was a tiny knob, or "button," which would later became his first rattle.

If he survived he might grow to be 6 feet long or more. But his chances of survival were small.

He was all alone. If he did not find his own food, he would starve. And if he did not recognize his enemies, they would kill him. Almost every creature that he met would attack him on sight or flee from him in terror.

Heat and cold might kill him too. Since he was "cold-blooded," his body temperature changed with the weather. The sun warmed him, but if he became too warm he would die of heat exhaustion. A sudden drop in temperature could paralyze him, and then he might freeze.

His glittering yellow eyes were rather weak. He could see shapes and movements several feet away, but beyond that his vision was blurred. And he could never close his unblinking eyes. They were protected by transparent shields, or "spectacles," which gave him a glassy stare.

He could not hear sounds as other animals do, since he had no external ears or eardrums. However he did have an inner ear. He was able to "hear" or "feel" through his skull bones, which picked up vibrations from the air and ground.

As he glided down the mountainside, he seemed helpless and weak. But he wasn't. He had powerful senses that other animals lack. And he possessed two deadly weapons that most animals feared.

His sense of smell was highly developed. He could smell not only with his nostrils but also with his dark forked tongue. As his tongue flicked out, it picked up tiny particles from the air or the ground or whatever it touched. Then it carried those particles back to his mouth. In the roof of his mouth were two pits, called a Jacobson's organ. These pits identified the odor of every particle his tongue picked up. His flicking tongue acted as a busy scout which would help him follow the trail of his prey.

He also had a pit on each side of his head, between his nostrils and his eyes. These pits looked like an extra pair of nostrils, but they were not used for smelling. Instead they were like sensitive thermometers, reacting to the slightest change in temperature. They told him when a warm-blooded animal was near and exactly where it was located. With these heat-sensing pits he could strike accurately at any warm-blooded prey.

His weapons were two pointed fangs. When he closed his mouth, the fangs folded back. When he opened his mouth, they sprang forward. The fangs were hollow. Each one was connected to a sac in his cheek which held a poisonous venom. When he struck at his prey, strong muscles would squeeze the venom through his fangs and into his victim's wound.

With his venomous fangs and heat-sensing pits, he belonged to the small group of poisonous snakes called pit vipers — rattlesnakes, copperheads and water moccasins. From the moment of his birth, he was able to inflict a painful wound.

And yet he was not aggressive by nature. In fact, he was rather timid. Unless he was cornered he would not try to fight his enemies. It was safer to disappear under a rock or into a hole. He would use his fangs only to kill the food he needed and to protect his life.

He had traveled less than 50 yards from his birthplace when he met his first enemy. A dark shadow appeared suddenly above him. He stopped short and lifted his head. Then he coiled up to defend himself.

A red-tailed hawk swooped down on him. It reached out with sharp talons, ready to carry him off and eat him. He hissed loudly, flicked out his tongue and lunged at the big bird. The hawk veered to avoid his head. As it swept past him, its talons brushed his back.

The hawk wheeled sharply and headed for him again. But he was already slithering into a hole beneath a rock.

He hid under the rock as it grew dark. After a while he felt the ground vibrating around him. The rock began to move. Dirt tumbled into the hole where he lay. The heat-sensing pits in his cheeks became warmer and warmer, and his flicking tongue carried strange scent-messages to his mouth.

Sharp claws were digging toward him. Quickly he began to crawl out of the hole. As he tried to escape, he came face to face with a hungry skunk.

The skunk seemed startled. It backed away and hesitated. Again the snake coiled and prepared to strike.

The skunk slapped the ground with both front paws. It jabbed at him and then jumped back. Now the rattler struck. But he could spring forward only a few inches — about half the length of his body. The skunk backed easily out of range.

Cautiously the skunk began to circle him. As it did, the rattler drew his body into S-shaped curves and squirmed rapidly toward a tangled pile of rocks. He squeezed in between two boulders just as the skunk tried to grab him.

He stayed there for the rest of the night. Early the next morning he crawled out and warmed himself in the sun. Then he continued his slow journey down the mountain.

That day he had his first meal. He was coiled up in a cool pile of leaves when a small fence lizard scurried down a nearby tree trunk. The lizard stopped to snap up a beetle. Then it circled the tree trunk and moved closer to the pile of leaves.

The rattler shot forward like a spring uncoiling. His jaws snapped open and his venomous fangs clicked into position. He stabbed deeply into the lizard's back.

Immediately he withdrew his fangs, coiled up and waited. The attack was over in less than a second.

The lizard looked around as if bewildered. It ran a few feet. Finally it staggered and fell.

Now the rattler crawled over to the dead lizard. With his

flicking tongue he examined the lizard's body. Then he opened his mouth wide and clamped down on the lizard's head.

Like any snake, he had to swallow his food whole. His backward-curving teeth were made for gripping, not for chewing.

First he pushed one side of his upper jaw forward as far as it would go. Then he pushed the other side of his upper jaw forward. After that he pushed forward with each half of his lower jaw. Special hinges allowed his lower jaw to drop down from the back of his mouth, so he could gape hugely. The two halves of his lower jaw were connected at the chin by an elastic muscle, which stretched like a rubber band. With his expandable jaws, he could swallow animals whose bodies were thicker than his.

Little by little he worked his jaws over the lizard's head and body. Glands in his mouth released saliva which coated the lizard and helped him swallow it. His skin was elastic too, and it stretched as he swallowed. By the time the lizard's tail had disappeared, the snake's slender body was bulging.

He crawled to a hollow tree stump and went inside. Then he stretched out to rest. Already he felt sluggish and drowsy. He did not want to move. He stayed there for several days, digesting his meal. Slowly the lump that was the lizard moved along the length of his body. The lump became smaller and smaller as his powerful digestive juices dissolved the lizard, bones and all.

Soon afterwards, something strange happened. His vision began to fail. A milky film formed over his eyes, and before long he was almost blind. Now he had to rely on his nostrils and his flicking tongue as he found his way about. He spent most of his time hiding in underbrush or beneath rocks and logs.

His skin had become dry and flaky. He had been growing fast, and by now he had outgrown his original skin covering. He was getting ready to cast off his outer skin.

A new layer of skin had been forming beneath the old one. His body had released a milky fluid between the two skin layers, causing them to separate. This fluid had covered his eyes too, because his transparent eye shields were part of his outer skin.

When the two layers of skin had separated, the milky fluid vanished. His eyes cleared, and he could see again. His old skin was now as thin as tissue paper.

He wanted to rub himself against any rough surface he could find. He crawled to a tree stump and kept pushing his jaws against it. The skin around his mouth began to crack and loosen. Gradually he pushed the old skin back across his head.

Then he slithered through the underbrush, rubbing against rocks and bushes, crawling out of his old skin. It peeled back from his head all the way down to his tail. At last he was free of it. His old skin lay on the ground like a long thin stocking turned inside out. The tiny ticks and mites that had burrowed beneath the scales were left behind.

Along with his bright new skin, he had also acquired his first rattle. It had developed from the small button at the tip of his tail. During his first year of life he would shed his skin four or five times, and with each shedding he would gain a new rattle. Each new segment would grow out of the previous one and would be loosely connected to it, causing a buzzing sound when he vibrated his tail.

If he lived to be three years old, he might have ten or eleven rattles. By then he would be about 2½ feet long and ready to mate. After that he would grow more slowly, shedding perhaps once a year.

In his fresh skin he set out again. That summer he roamed about a mile from his birthplace. For a few weeks he lived in an abandoned woodchuck burrow beneath a big boulder at the edge of the woods. He came out to prowl early in the morning and again at dusk. When the midday sun grew too hot for comfort, he retreated back into the cool darkness of the burrow.

He hunted mostly for lizards and small field mice. As he grew older he would eat larger animals — rats, gophers, squirrels and rabbits. But he was not yet big enough to squeeze them into his elastic body.

He wasn't fast enough to chase his prey, so he had to hide and wait until a meal came his way. Often he would lie in ambush for hours, ready to strike at the right moment. When he was lucky, he gorged himself. When he wasn't he simply waited for another chance. If necessary, he could go days or weeks without food.

Every so often he crawled down to a nearby brook. Then he would thrust his mouth into the cool water and suck it down his throat. One evening as he was drinking, a deer almost stumbled over him. The big animal snorted and reared up on its hind legs. Aiming for his head, it tried to crush him with its sharp front hooves. He squirmed violently out of the way and plunged into the brook. He had never been in the water before, but he was a good swimmer.

The next day two hikers came upon him as he lay coiled near the entrance to his burrow. They approached cautiously, not sure at first what kind of snake he was. He vibrated his tail, but since he had only one rattle so far, he could not give a warning buzz.

Raising his head to face them, he flicked out his tongue and hissed. Suddenly one of the hikers picked up a rock and hurled it at him. At that instant he lunged forward to strike, and the rock barely missed his head. The hikers had jumped out of range. As one of them grabbed a stick, he retreated back into his burrow.

Toward the end of September he felt a strong urge to travel back up the mountainside. Crawling through tangled grass and underbrush, resting now and then beneath stones and logs, he returned the same way he had come. Finally he reached the rocky ridge where he had been born.

Many other rattlesnakes had gathered here. Just below the ridge, a small opening between the rocks led to a cave hidden deep inside the mountain. The cave was a snakes' den. It had been inhabited by generations of rattlers and other snakes.

Dozens of snakes were scattered all along the mountain ridge. Some were 5 or 6 feet long and as thick as a man's arm. He crawled onto the ridge and rested among them. Perhaps his mother was there, or some of his brothers and sisters. If so, he didn't know them.

As autumn arrived he remained at this rocky outpost near

his ancestral den. The chilly nights made him tired. He began to spend most of his time in holes and crevices. He came out only on warm days to soak his body in the bright October sunshine.

When the weather grew colder he retreated into the mountainside den, which was buried several feet below ground level. Other rattlers crawled into the den with him, along with some copperheads, milk snakes and garter snakes. They coiled and twined their bodies together for extra warmth, forming a great writhing ball.

Now he felt more and more sluggish. The den was so dark that he could not see the snakes next to him. The smells and vibrations around him faded. He no longer felt the increasing coldness of his own body. As he drifted off, he was no longer aware of anything.

His breathing slowed. His heart was barely pumping. His body was motionless. He would stay that way for the rest of the winter, hibernating with the others. Outside the temperature might drop to zero, but the den would stay warm enough to keep the snakes alive.

With the rising temperatures of spring, he would leave the den and venture out into the world again. His second summer would be very much like his first. He had already experienced almost everything that a timber rattlesnake can.

Bald Eagles

She was hungry. Her parents had stopped bringing food to the nest, and for two days she hadn't eaten.

Now she felt restless and irritable. She opened her wings, flapped them noisily and rose several feet into the air. Then she dropped back down to the floor of the nest.

Her brother chirped and jumped to one side. They were the same age, nearly three months old, but she was bigger than he and bolder too. He tried to keep out of her way.

Soon they saw their mother's white head gleaming in the sun as she flew toward them over the treetops. Clutched in her talons was a large fish. Excited, the two eaglets hopped up on the edge of the nest and chirped loudly, calling out to their mother.

She circled above them, calling back with throaty

cackles — "carruck, carruck, carruck." But she didn't land in the nest. Instead she flew twice around it. She settled in a nearby tree where she ate the fish herself.

Holding the fish with her talons, she tore off chunks of flesh with her curved yellow beak. Each time she grabbed a piece, she lifted her head and gulped it down. The hungry eaglets leaned over the edge of the nest, crying and complaining as they watched their mother eat.

When she finished, she picked bits of food from her scaly yellow legs and toes. Then she cleaned her beak by rubbing it against a branch. After that she flew away.

The male eaglet hopped back into the nest. He pounced on a twig, jabbed it with his beak and tossed it into the air. Then he fluffed his feathers and started to preen them. His sister remained perched on the edge of the nest, staring at the wilderness valley far below.

The nest, or eyrie, was a massive platform of branches and sticks perched high atop a cottonwood tree, 100 feet above the ground. It had been built years before by the eaglets' parents, who were lifetime mates. In this remote valley, far from the dangers of civilization, they had raised several broods of healthy young.

Every spring the adults returned to their nest and repaired it, adding more branches and sticks. And every year the nest grew larger. Now it was taller than a standing man and just as wide across, with walls about 2 feet thick. Pine needles, feathers and down padded its spacious floor, where the eaglets were growing up.

They had hatched in May. As small helpless infants they

had been guarded closely by their parents, who protected them from hawks, crows and other hunting birds. By July the eaglets had outgrown these high-flying enemies. By August they were almost as big as their parents. But they still couldn't fly. They had never left their gigantic treetop nest.

Several times that morning their parents had circled the nest with food in their claws, trying to coax the eaglets out. Now the young female leaned forward from her perch, watching the river that rushed through the valley below. With her sharp eagle's eyes she could see fish leap from the water, then disappear in the swirling rapids.

She leaned forward a bit more. The wind rippled her dark brown feathers, and her fierce eyes glowed. Suddenly she jumped. Her feet left the nest. She stretched her neck, opened her wings and flapped them hard. She was flying through the air. For the first time in her life, the eagle was flying!

A rising current of air swept her over the treetops, carrying her higher and higher. Now she hardly moved her wings at all as she glided easily across the river and over a grassy meadow beyond.

Then she began to lose altitude. She found herself sinking, and again she stroked her wings. She zigzagged up, then veered sharply down. The ground came closer and closer, rushing by beneath her. A strong wind was pushing from behind. Pumping her wings frantically, she bumped into the meadow and tumbled head over claw through the grass.

Where was she? She scrambled to her feet and looked around. In the distance, on the other side of the river, she saw her nest. Overhead, circling and calling, she saw her parents. She shook her body, fluffed her feathers and settled down on her stomach to rest.

She sat there for quite some time, poking about in the grass. With her long curved talons and her sharply hooked beak, she could defend herself against all but the biggest animals. Flying, however, was her best defense, and for the moment she was grounded. Her parents stayed nearby, perched in a tree where they stood guard.

After a while she stood up and tried to take off. She flapped her wings and jumped, but she rose only a few feet. She tried again and still dropped back to the ground. Finally she began to run. She ran across the meadow flapping hopefully, and this time her wings carried her into the air.

Heading home, she made short, uncertain hops from tree to tree. Along the way she rested on branches. When she finally touched down in the nest, her wings drooped heavily at her sides.

Her brother was gone. He had flown as far as a neighboring tree, where he sat calling from a branch.

From then on she and her brother left their nest every day. As they practiced flying, their muscles grew stronger and their flights longer. Gradually they learned to steer by using their fan-shaped tails as rudders and by adjusting the flight feathers on their broad sweeping wings. They learned to land against the wind, pointing their tails down and pulling their wings back as brakes. And they learned to catch long, lazy rides on the warm air currents that drifted upward from the ground.

Soon they were wheeling and soaring high above the

valley. They would glide slowly in wide circles, coast down-
wards, find another air current and rise again until they became small dark dots against the sun.

At first they stayed close to their parents, flying just
behind them. While they were nearly as big as the adults, they
didn't look like them. Their parents had pure white heads and
tails, while the rest of their plumage was dark brown. The
young eagles were brown all over, except for whitish patches
on their wings. Their eyes and beaks were dark instead of
yellow. They would not develop the distinctive white heads of
mature bald eagles until they were four or five years old.

When they weren't flying, they spent most of their time
at their treetop nest. Once again their parents brought food,
but already the young eagles were learning to hunt.

One morning the young female followed her father to a
lookout post beside the river. The two of them sat side by side
on a branch of a tall tree. From their lofty perch they watched
fish leaping and darting in a shallow pool at the river's edge.

Suddenly the father eagle dropped from his perch. He
opened his wings, swooped down toward the river and turned
sharply. Water splashed up against his belly and wings as he
skimmed over the surface, reaching down into the water with
his talons. When he flew up again, he clutched a wriggling
fish.

The young eagle had followed her father down from the
tree, sweeping past him over the river. Now she wheeled and
turned and flew quickly to his side. As she landed beside him,
she crowded in anxiously to snatch pieces of the kill.

She tried fishing herself but at first had little luck. The
fish moved swiftly and then disappeared in the currents. As

she swooped down on them, she had to check her speed and strike the surface with perfect timing. Otherwise she grabbed nothing but clawfuls of water. Sometimes she hit the water with such force that she tumbled over and drenched herself before taking off again. It was easier to pick up dead fish along the shore. That's what she did whenever she could find one.

While she preferred fish, she also hunted other prey. She and her brother learned to fly low over the river marshes, searching for wild ducks among bushes and reeds. They had to be fast when they swooped to attack. The ducks often escaped by dashing into deep water.

When she caught a duck she crushed it with her talons, killing it quickly as she soared back into the sky. She carried it to a nearby plucking post, held it firmly with one foot and pulled out the feathers with her beak. Then she carried the plucked duck back to the nest and ate it there. When she had picked the bones clean, she removed them from the nest and dropped them far away.

At times the young eagles followed their parents beyond the river, scouting meadows and mountains for squirrels, rabbits and other small animals. Their vision was about eight times sharper than a human's, and as they glided silently through the sky they could spot a rabbit from a mile away. They learned to watch and wait, hovering in the sky until their prey was out in the open. Then they would fold their wings and plummet toward earth as the frightened animal ran for cover.

Gradually the youngsters became expert hunters, fast enough to attack ducks in flight and snatch a meal out of the air. They also learned to steal food from the ospreys, or fish hawks, that invaded their own stretch of the river. The ospreys often had better luck than the eagles, because they could dive underwater while the eagles could not.

One morning the eagles watched an osprey rise from the river with a trout in its claws. A moment later their mother flew off in pursuit. With rasping screams — "screeeee-ack-ack-ack" — she closed in on the smaller bird. The osprey cried out and turned sharply. But the eagle followed. She overtook the osprey, dove beneath it and began to drive it higher into the sky.

Suddenly she twisted in midair, flipping over on her back. Still shrieking, she beat at the osprey with her giant wings and slashed at it with her talons and beak. The osprey cried out again and finally dropped the fish. As the fish fell, the eagle wheeled and swooped, diving down to catch it.

The eagles usually spent the morning hunting, satisfying their appetites by noon. If the day was dark and overcast, or if there were no winds, they might spend all afternoon perched silently in trees. But in good flying weather they rode air currents for hours until it was time to roost.

With wings spread wide they soared high above the valley floor, dipping and turning, sailing along with the winds. Suddenly one of them would fold its wings and plunge toward earth like a meteor. Then the eagle would swoop up again, roll over in an aerial somersault and streak through the clear blue sky.

At dusk they returned to their roosts to settle down for the night. The young eagles no longer slept in their nest. Instead they perched on branches. Standing on one leg, with their heads buried in their shoulder feathers, they slept until dawn. Like most bald eagles, they often snored.

As autumn arrived, the youngsters spent less and less time with their parents. Now they hunted for themselves, catching all their own food. The adults began to fly upriver alone, leaving the young eagles behind. Often they didn't return for two or three days.

By the middle of October, the youngsters were nearly six months old. Aspens and cottonwoods along the river had turned to brilliant shades of red and gold. Ducks were banding together in large flocks, getting ready to fly south. Soon the wild cries of Canada geese rang over the valley as the migrating birds passed overhead in great wavering V's.

One morning, when the young female woke up, her brother was gone. Her parents had flown upriver several days earlier and hadn't returned. No other birds were in sight. A light snow had fallen during the night, dusting the trees and ground with white. Thin crusts of ice glittered along the edges of the river.

The eagle called loudly to greet the morning. She shook herself, fluffed her feathers and combed them carefully with her beak. When she had finished her preening, she dropped from her perch, spread her wings and flew upward into the sky.

She flew over the treetops, past the nest where she had hatched and across the river she knew so well. When she had climbed high above the valley floor, she banked and made a wide turn. Then she headed south, following the path of the river. Her strong wings carried her toward the unknown mountains in the distance beyond.

Beaver Kits

The two yearlings had just entered the lodge when their mother turned on them. She hissed at them and snarled, baring her big orange teeth. Then she chased them down the plunge hole and out into the pond beyond.

With mumbles and murmurs of complaint, they swam back and forth in front of the lodge. Soon their mother chased their father out too. Bewildered, they looked to him for protection.

He led them to an underground bank den at the far end of the pond. Normally the beavers used this den as a temporary resting or hiding place. Now it became a new home for the three outcasts. All of them crowded in together.

For the first time in their lives, the yearling kits slept outside the lodge where they were born.

After that they watched their mother from a distance. She kept swimming to shore, where she gnawed branches from birch and aspen trees. Then she carried the branches home and shredded them with her razor-sharp teeth. She used the long soft fibers to cover the beavers' sleeping shelves. As the fat old female went about her business, she gradually changed all the bedding in the lodge.

The next day she disappeared inside. That night the yearlings heard small stirrings and strange mewing sounds. And they could smell the sweet musky odors of newborn kits.

Soon they saw their mother out and about again. She would swim to shore, comb the water out of her fur and feed hungrily on leaves and bark. Then the shrill squeaking cries of the infants would send her hurrying back to the lodge.

Early one evening in May, the yearlings saw the new kits for the first time. Three wide-eyed infants came swooshing out of the lodge behind their mother. They weighed little more than a pound apiece, yet they swam just as naturally as they breathed. Holding clenched paws like little fists against their chests, they paddled with their webbed hind feet and steered with their flat scaly tails.

From then on the baby kits came out regularly at dusk and again early in the morning. They swam one after another in slow circles, sniffing at every leaf or sliver of bark that floated by. Slipping and splashing, they climbed the sloping walls of the lodge to gnaw at sticks and tumble each other about. When their mother joined them, they pulled and tugged at her until she nursed them or groomed their soft baby fur.

Before long the yearlings approached the infants and

nuzzled them curiously. Then they began to shadow the babies — swimming behind them, diving beneath them, splashing about in front of them. When the new kits were about two weeks old, the family was reunited. The yearlings and their father left their bank den and moved back into the family lodge.

The lodge was a marvel of animal construction. It rose from the middle of the beaver pond like a tall island of sticks and mud. The beavers went in and out by means of two water-filled tunnels. The entrances to these tunnels were hidden near the bottom of the pond. Each tunnel curved upward to a circular living chamber, which stood a few inches above water level. Raised above the floor of this small dark room were sleeping shelves covered with soft shredded wood.

The beavers slept during the day and were active mainly at night. Sometimes they came out to bask in the afternoon sun, but they felt safer under cover of darkness.

They always left the lodge at dusk. The impatient yearlings usually led the way, diving into the pond with a muffled splash and popping up to the surface a few feet away from the lodge. At first they swam cautiously back and forth. Keeping their heads above water and working their round rubbery noses, they tested the air for telltale signs of danger. They had sharp ears, and they could pick up a menacing scent from a great distance.

If nothing alarmed them they played for a while, chasing each other across the pond or following one another in endless circles. Meanwhile their parents came shooting out of the lodge like furry torpedoes, with the baby kits trailing close behind.

When the yearlings waddled ashore, they spent ten or fifteen minutes drying and grooming their fur. Using their broad flat tails as props, they would sit upright on the bank of the pond. With their front paws they would rub their eyes, shake the water out of their ears, comb their whiskers and give their bellies a long satisfying scratch. Then they would squeeze the water out of their fur, using the split toenail on each hind foot. Finally they would reach down to the oil glands beneath their tails and rub the oil through their long outer fur, keeping it sleek and waterproof. Their underfur was softer and thicker and helped keep them warm.

After grooming it was time to eat. With their self-sharpening, chisel-shaped teeth, the yearlings cut branches from bushes and trees. Holding the branches with both front paws, they nibbled daintily at the tender green buds and leaves. Then they peeled off the sweet bark as if they were gnawing corn on the cob. Small twigs could be eaten whole, like crisp celery stalks.

Bark and twigs were their main food, but the springtime forest offered many seasonal delicacies. The yearlings feasted on ferns and grasses along the banks of the pond. They dug up roots, wiped off the mud and ate them whole. And every night they searched the bottom of the pond for the juicy shoots of water lilies.

While the yearlings and their parents fed near shore, the infants at first stayed close to home. They spent most of their time wandering in and out of the lodge entrance. In a few weeks the baby kits grew bolder, and they began to follow their mother on her nightly rounds. They still depended on her rich yellow milk, but now they sampled different kinds of beaver food. Churring and chattering with excitement, they would crowd around their mother and nibble on the same branch she was eating.

She always watched the infants closely, especially on land. Sometimes she picked up a kit and carried it down to the water the same way she carried branches — supported on her front paws and held in place by her resting chin.

The yearlings watched out for themselves. In the pond they felt confident, and they swam about freely. But on land they were always cautious. As they lumbered about foraging food, they kept an uneasy lookout for enemies. Any large meat-eating animal was a threat. Any strange smell or sudden sound was a danger signal.

One summer evening, as they nibbled on mushrooms away from shore, they heard a sharp sudden WHACK! Their father had smacked the ground with his flat tail. In an instant the yearlings took off. Crashing through the underbrush, whacking the ground with every leap, they raced for the safety of the water. As they flung themselves into the pond, each one gave the water a resounding tail-smack. Then they disappeared.

A few moments later, a bobcat emerged from the woods and looked around. He walked to the edge of the pond, crouched flat and lapped its cool water. The beavers were out of sight. All of them were safe inside their island-lodge.

The yearlings knew from experience that tail-smacking meant D A N G E R! When they heard that sound they dashed for the pond. The baby kits did not always react so quickly. Often they ran with the others, but just as often they seemed confused. Then they would run around in circles or pause to sniff and listen until their mother came rushing over to fetch them. They had not yet learned how to recognize their enemies.

The kits often played for hours on end, and sometimes the yearlings joined them. They would chase the young ones back and forth in wild games of water tag, or lead them through twisting mazes of tangled underwater brush, or tussle gently with them in the shallows near shore. Occasionally the whole family tumbled about together. They would gather around an old log and gnaw the rotten wood to shreds. Then all of them would roll on their backs in the dry wood powder.

Summer was an easy time for the beavers. There was little work to be done and food was plentiful. Their favorite summertime food was the algae, or green pond scum, that floated on the surface of the pond. With their paws they would pull in great swaths of algae and slurp it up like green spaghetti.

Every day the old beavers inspected their dam at one end of the pond, puttering around and making small repairs on the spot. But mostly during the summer months, the family loafed and ate and grew fat.

With the first brisk days of autumn, this lazy existence came to an end. The beavers sensed the approach of winter. They began to prepare for the cold months ahead.

Every night after filling their bellies, they cut down trees along the shores of their pond. The adults were fast, tireless workers. The yearlings worked almost as hard, cutting and hauling smaller pieces of timber.

A yearling would stand upright before a tree with its tail braced firmly on the ground. Grasping the trunk with its paws, it would bite out chips of wood and gnaw its way around the trunk. As the tree began to creak and sway, the yearling would dash for the pond, dive under and wait until the tree fell.

Then it would return to cut up the timber, biting off branches and gnawing the trunk into small logs. After that it would push, drag or carry the timber down to the pond.

To begin with the yearlings and their parents used this timber to reinforce their dam. They strengthened the dam and raised it so the pond would be deep enough to keep from freezing to the bottom.

Working with paws and teeth, they shoved sticks and logs into the walls of the dam. They stuffed holes and plugged leaks. Wheezing and grunting, they piled fresh timber on top of the dam and dumped heavy rocks on top of the timber. Then they plastered the dam with armloads of mud scooped up from the bottom of the pond. As the mud seeped in among the sticks, it helped seal the dam and make it watertight.

When the dam was ready, the beavers floated timber out to their island-lodge. They reinforced the lodge as they had the dam, shoving new sticks into the walls and plastering them with armloads of mud.

Each beaver labored alone, without paying much attention to what the others were doing. Sometimes one would stop to watch another at work. Then it would go about its own business.

The baby kits seemed excited by all this activity, but they weren't much help. They followed the others everywhere — tugging at branches, chewing at sticks, poking into the mud and getting in the way. Once in a while one of the kits would grab a stick and try to wedge it into place. Then the youngster might yank the stick loose, carry it back to the water and play with it.

After repairing their dam and lodge, the yearlings and their parents began to store wood for their winter food supply. Now they worked through the night as they cut down trees and hauled timber to a storage pile beside their lodge. During the frozen winter months they would eat nothing but the bark of these trees.

They preferred aspens, so they spent most of their time cutting those. Diving to the bottom of the pond, they pushed sticks and branches into the mud. Then they wedged more sticks and branches into the growing storage pile. Eventually the pile of food-wood rose above the water like a small lodge next door to the big one.

Winter came suddenly. Ice hardened across the surface of the pond. Snow covered the walls of the lodge, and the mud that held the walls together froze solid. Inside the darkness of the lodge, the small living chamber was warm and cozy because of heat given off by the beavers' bodies.

The beavers had not plastered the roof of the lodge. Instead they had left small ventilation holes between the tangled layers of crisscrossed sticks. As the warm air inside drifted upward and escaped through the roof, it formed vapors of steam which looked like smoke rising from a chimney.

Although the surface of the pond was frozen, the underwater doorways to the lodge remained ice free. When the beavers were hungry they could slip into the water, swim beneath the ice crust and grab a tasty branch from their submerged food pile.

Every so often they heard an intruder outside their walls. As hungry animals prowled the winter forest, they were attracted to the snow-covered lodge with steam rising from its roof. They could smell the tantalizing odors of the plump beavers inside. One afternoon a late-hibernating bear walked out across the ice. He sniffed at the lodge, chewed at its walls and tried to claw his way inside. But the beavers were locked safely behind the walls of their frozen fortress.

Only one enemy was clever enough to threaten them. On a cold morning in February, the beavers were awakened by the sounds of ice chipping and cracking. The old male left the lodge to investigate. He slipped into the plunge hole and swam beneath the ice in the direction of the dam.

The yearlings and the young kits waited with their mother. A few minutes later they heard the sounds of frantic splashing and struggling. Suddenly the old male came swimming up through the entrance tunnel, leaving a trail of blood behind him. His left hind foot was ripped open, and he had lost a toe. Whimpering and trembling, he limped to his sleeping shelf and nursed his wounds.

At the bottom of the pond, near the base of the dam, he had blundered into a steel trap. Hunters searching for beaver skins had attached the trap to a long pole, pushing it under the ice. When the trap clamped down on the old male's foot he had twisted and squirmed, gasping for air. He had managed to free himself before drowning. His foot was mangled, but he was lucky to be alive.

When the spring thaw finally arrived a few weeks later, his wounds had almost healed. Now the beavers heard ice creaking and shifting as it began to break and melt. For the first time in months they left their lodge, popping up to the surface of the pond and venturing into the rain-drenched forest.

Soon the first ducks came flying in over the treetops. Frogs and salamanders wriggled out of the mud at the bottom of the pond. Skunk cabbages pushed up through the soil of the forest floor. After feeding on bark all winter, the beavers greedily devoured the newly sprouted shoots.

The younger kits were yearlings now. And the two-year-olds were almost mature. They were old enough to leave home. But they did not leave willingly.

Early one morning in April, the biggest two-year-old sat quietly on a landing ramp outside the lodge. Suddenly his father attacked him. With no warning at all, the snarling old male dashed forward and chased his son into the water.

Churring and chattering with alarm, the young beaver scrambled into the lodge. His father pursued him, slashing angrily at his back. When the two-year-old ran over to his mother, she swung around and hissed.

Driven out of the lodge, the young beaver raced across the pond and climbed ashore. He began to groom his fur as though nothing had happened. For a long time he dawdled on the bank, sniffing at the ground, nibbling at sticks, scratching himself nervously. Finally he plunged back into the water.

His father was waiting. The old beaver rushed forward and attacked his son again with tooth and claw. He did not try to draw blood, but it was clear now that he meant business. Terrified, the two-year-old broke away and swam for his life.

His father chased him only as far as the dam. He stopped there, pacing back and forth along the crest of the dam as the youngster headed downstream and disappeared around a bend. After that the old male went back to the lodge and drove out the other two-year-old.

Like all beavers the same age, the two-year-olds left their birthplace for good. Their mother was about to deliver a new litter of kits, and the colony could not support so many beavers. Only a limited amount of food grew around the pond.

Traveling separately, the outcasts moved cautiously downstream through strange, unknown country. At first they were fearful and confused. But as they journeyed on their fears faded, and a growing restlessness stirred within them.

Now they were independent. It was time for them to seek mates and start new beaver colonies of their own.

Lion Cubs

He crouched low in the grass, every muscle poised and tense. His tail flicked back and forth, and his bronze eyes glinted in the morning sunlight.

For a moment he waited. Then he crept forward, scarcely breathing. His belly brushed the ground as he inched closer to his unsuspecting target.

Suddenly he sprang through the air — reaching out with anxious claws. He landed on his sister's back, bit fiercely into her neck and pulled her down to the ground.

The two cubs grappled and wrestled, swatting each other with their big paws. Then they scrambled to their feet and raced wildly toward a thicket.

Their mother raised her sleepy head and watched the

cubs tear past. She yawned widely, rolled over and dozed off again.

All around her, lions were sprawled out in the shade of a flat-topped acacia tree. They were crowded close together — some on their sides, others on their backs — with their legs and bodies touching. One lioness had climbed up into the tree where she lay draped across a branch. Her chin rested on the branch, and her arms dangled toward the ground.

From their resting place the lions looked out at the shimmering African plain with its dark clumps of thorn and acacia trees. This plain was their hunting ground, the territory claimed by members of their pride. Other lions were not welcome here.

A few weeks earlier, during the dry season, the plain before them had been parched and bare. Most grazing animals had gone elsewhere in search of food. With few prey left to hunt, the lions had grown lean and hungry. Two cubs belonging to the pride had starved.

But now the rains had returned. Rivers were swollen, grass grew high and the migrating herds had surged back onto the plain. In the distance, wildebeest with curved horns and white beards grunted and snorted as they munched endlessly on the grass. Zebra galloped across the horizon, braying loudly. Gazelle went leaping and bounding through the haze. For the lions, the rainy season was a time of plenty.

The pride sometimes claimed twenty members or more, depending on the comings and goings of the males and the birth of cubs. On this particular morning, eleven lions had gathered beneath the acacia tree — two massive males with full manes, four sleek and tawny females and five frisky cubs about six months old.

The female cubs would stay together throughout their lives. The males would leave the pride when they were about three years old and just growing their long silky manes. They would become wandering nomads like their fathers — roaming through plain and woodland, drifting from one pride to another. Until then, they were all members of the same family.

As the adults rested, the cubs played. One cub pounced on a twig. She clamped down with her teeth, shook her head violently and dragged the twig across the ground. Then she grabbed it with both paws and rolled over on her back. Clasping the twig close to her chest, she kicked her feet in the air.

Another cub sat beside a male lion and watched the flicking black tassel at the tip of his tail. Suddenly the cub grabbed for the tassel. He caught it between his paws and started to chew on it. The big male looked up and snarled, but the cub would not leave him alone. He ran over to the lion's head and tugged at his tangled brown mane. The lion was not in a playful mood. He rose to his feet, shook the cub off and cuffed him sharply.

The cub bounded over to a sleeping female. He jumped up against her, draped himself over her face and pulled at her paws. At first she tried to push him off. Then she rolled over and covered him with her body until he managed to squirm free.

Before long the cubs exhausted themselves. One by one they squeezed in among the adults, flopped down beside them and fell asleep. For the rest of the morning the lions did nothing but nap.

Around noon the cubs became restless again. They got up and wandered aimlessly about. Then all five of them ran over to the nearest lioness to nurse.

Snarling and shoving, they fought over the four teats of the resting female. Annoyed, she turned her head to warn them, growling and baring her sharp yellow teeth. But she did not try to push them off.

One cub was finally elbowed aside by the others. He ran over to another lioness. She rolled over on her back, lifted her hind leg into the air and let the cub suckle by himself.

For a while the cubs played again — stalking, rushing, chasing and wrestling. But they spent most of the hot, humid afternoon cuddled against the sleeping adults. Every so often one of the lions would change position. It would stand up, move slowly to another resting place, then collapse back into the grass with a heavy sigh. As usual the lions relaxed all day. They rarely exerted themselves before evening.

At dusk they finally began to stir. They yawned and stretched and yawned again. Lionesses rubbed their heads and bodies against each other, making soft humming sounds in their throats. All of them milled about for a few minutes, as though they were tense and expectant. Then the adults moved off together, leaving the cubs behind.

The cubs sat in the grass beneath the acacia tree. They watched the adults walk toward a low hill, but they did not attempt to follow.

As the light faded, the lions stood silently at the crest of the hill. Purple shadows fell over the plain. Slowly it grew

dark. From the distance came the grunting sounds of wildebeest and the occasional bray of a zebra.

Finally the lions vanished into the night.

The cubs waited quietly for nearly two hours. At last
their mothers came trotting toward them out of the darkness.
The youngsters ran to greet the two females, rubbing against
them and licking their mouths. The lionesses made low
moaning sounds, which meant "Come." Then they turned to
lead the cubs back to the kill.

Although the cubs still nursed frequently, they were too
old to live on lions' milk alone. They needed meat to survive.

The hungry youngsters bounded eagerly ahead of their
mothers. Now and then they stopped to look back and wait
until the females caught up and showed them the way. They
traveled about a mile and a half before they reached the
others.

By the time they arrived, the carcass of a zebra had been
torn to pieces. Most of the meat was already gone. Each lion
crouched by itself, gnawing on a bone. The biggest male had
taken possession of the entire rib cage and head. He lay on
his belly beside it, guarding his prize. When any other lion
came too close, he raised his head and growled a warning.

But he did not threaten the cubs. As they rushed over to
join him, the big male glanced briefly at them. Then he went
back to his bone. The cubs crowded in beside him, squabbling
among themselves for a share of the meal.

Soon the last bones had been picked clean. One by one
the cubs wandered away from the kill and sought out their
mothers. They had just gorged themselves with meat, but
even so, they all began to suckle.

The lions rested by the remains of the zebra as a full moon rose in the African sky. Long after midnight they finally moved on. This time the cubs went along.

The four lionesses stayed in front of the group, advancing cautiously through the tall grass. The cubs tagged playfully behind, pouncing on sticks and small scurrying insects, then bounding ahead again. The two males brought up the rear. They walked along slowly and deliberately, their shaggy heads nodding with each step.

The males seldom hunted for themselves. They almost always lagged behind and let the females make the kill. It was just as well. While they were bigger and stronger than the females, they were not as swift or agile. With their heavier frames and flowing manes, they were more likely to be spotted by alert prey. By keeping to the rear the males could help guard the cubs, who were still small enough to be attacked by leopards and hyenas.

Every few minutes the females paused to look and listen. As they crossed over a low ridge, they spotted a small herd of wildebeest about a quarter-mile away. Swiftly the lionesses spread out. Then they sat in the grass and waited. Only their heads were visible in the moonlight.

Behind them the cubs and males waited too. The cubs would not be big enough to take part in the actual kill until they were nearly a year old. They would not be able to hunt large prey by themselves until they were about two. But they were already learning. As they sat beside the males, they watched every move the females made.

Gradually the grazing animals drifted closer. A breeze was blowing across the plain, but the lions were downwind and the wildebeest did not pick up their scent. They were not yet aware of danger.

For an hour or more the lionesses remained concealed in the tall grass. Finally they fanned out. Two of them crept toward one side of the herd and two toward the other side. Crouching low they inched closer, their eyes fixed on the shadowy forms up ahead, their tense bodies straining forward. One wildebeest glanced up and peered through the darkness toward a stalking lioness. She stopped dead in her tracks with a paw suspended in midair.

When the females were nearly within striking distance, they all crouched motionless. No signals passed between them, but they had hunted so often together that they knew how to work as a group. Three of them stayed where they were. The fourth began to backtrack. She moved silently to the rear, melting into the darkness.

The lioness circled far around the unsuspecting herd. Then she approached boldly from the opposite direction. She came directly toward the wildebeest as the other lionesses lay flat on the ground, completely hidden in the grass.

Suddenly the wildebeest stopped their incessant grunting. They had scented lions. Panic swept through the herd as the animals bolted and scattered in all directions.

One of them plunged headlong toward a hidden lioness. As she reared up, the wildebeest twisted and turned, trying desperately to escape. The lioness lunged from behind, grabbing the animal's rump with her sharp hooked claws. Another lioness sprang out of the darkness at the wildebeest's shoulder. Together they pulled the struggling animal down to the ground. A powerful bite on the throat strangled the wildebeest and ended its life.

As the animal fell, the other lions came roaring out of the night. With slashing claws and snapping teeth they flung themselves onto the carcass. Snarling and cuffing, each lion fiercely defended its share from the others as they tore off chunks of meat and gulped them down. The cubs shoved in among the adults, fighting for their share too.

One cub tried to pull a piece of meat away from a lioness. With a menacing growl and a swift cuff of her paw, she knocked the youngster backwards. He went sprawling into the grass. But he bounded up again and nosed his way back into the writhing mass of lions crowded flank to flank around the wildebeest's body.

The lions ate until they could eat no more. For a long time afterwards they lingered near the remains, gnawing halfheartedly on bones, then looking up and growling softly. It was almost dawn when they finally lifted themselves off the ground and trudged heavily away.

Even before they were out of sight, other animals moved in to claim what was left of the spoils. Several hyenas had been lurking in the darkness, circling the kill and waiting. Now they rushed over to grab bones and carry them off. Two jackals slipped cautiously up to the remains, snipped off slivers of meat and retreated quickly. Vultures dropped from the sky and settled on the carcass to pick it clean. The lions paused once to look back and growl at the scavengers. Then they went on their way, too glutted to care.

At daybreak they were all sprawled out beneath the

same acacia tree. As the sun rose, the lions stirred. They stood up and looked out at the grassy plain that was their private hunting ground. A low moaning sound broke the early morning stillness as one of the males opened his cavernous mouth. Then a great roar welled up from deep inside his throat.

Now the other adults joined him. Lions and lionesses roared together as a pride, greeting the day with a thundering chorus of roars that echoed across the plain and challenged all outsiders.

The cubs sat silently beside the roaring adults. One cub opened his mouth and tried to roar with them. He worked his throat muscles, but all he could muster was a soft "ooo." Before he could roar like a lion, he would have to grow up.

Gorilla

She was the first to awake. Birds were singing everywhere, and a bright sun was rising over the African rain forest.

She sat up in her nest, blinked her eyes, stretched her arms and yawned. Then she looked over at her mother and infant brother, sleeping soundly in their own nest a few feet away. Mother lay on her side with her shaggy arms wrapped around the baby.

He was six months old. He still nursed at mother's breast, rode on her back and shared her nest at night. Sometimes his sister crowded into the nest with them. But mostly she slept alone.

She was four years old — an awkward age for a wild gorilla. She was too old to behave like a baby and too young to be treated like a grown-up ape. She was just about half-grown. She would not be mature until the age of seven or eight.

As she sat in her bed at the foot of a tree, she reached out and grabbed a stalk of wild celery. With a quick jerk she yanked it from the ground. Holding the celery with both hands, she bit off the tough outer bark and ate the juicy center of the stalk.

Soon the other gorillas began to wake up. One old female sat up and scratched her ear. For a moment she stared down at her colossal belly. Then she curled up and fell asleep again.

A big male raised his head and gazed drowsily into space. He stretched a hairy arm to one side and yawned widely. His teeth were covered with black tartar. Finally he climbed out of his nest and wandered about on all fours, walking on the knuckles of his hands and the soles of his feet.

The gorillas were slow risers. Nearly an hour passed before all of them had left their beds. Gradually they spread out through the forest for their morning feeding.

Food was everywhere. As strict vegetarians, they ate many kinds of green plants and fruits. Some of them sat on the ground, stuffing wads of greenery into their mouths, then reaching out for more. Others wandered from bush to bush, picking and choosing as they went along. As they ate they smacked their lips, grunted with pleasure and frequently belched.

The four-year-old female sat near her mother and baby brother. She reached for a shrub, snapped off a branch and splintered it with her teeth. Then she chewed on the tender white pith inside. When she finished that, she pulled the top of the shrub close to her face. With her thumb and forefinger she plucked off its purple blossoms and popped them one by one into her mouth.

After a while she stopped eating and rolled over on her back. The morning sun warmed her tight round belly. Then she shifted to her side and grabbed the sole of her right foot with her right hand.

Finally she stood up and ambled over to a tall bush. She climbed to the top of the bush and squatted there, swaying back and forth as she watched the others eat.

The gorillas fed for about two hours. Toward the middle of the morning they stopped foraging and gathered together in a clearing. It was time for their daily rest period.

The leader of the group sat at the base of a tree and leaned back aganst the trunk. He weighed perhaps 450 pounds — more than twice as much as any full-grown female. And when he rose on his short, bowed legs, he stood about 6 feet tall. Except for two large silvery patches on his back, his hair was a glossy blue-black. His dark face shone like polished marble. A massive ridge ran across his brow and hung over his eyes, which were dark brown and surprisingly gentle.

The others seemed to enjoy his company. They crowded around him, lying on their backs, their bellies or their sides. About fifteen of them lived together in the forest. Most had been members of this family group all their lives.

Some of the adults fell asleep in the sun. Small infants sat nursing in their mothers' arms or played on the ground nearby. Older youngsters wandered freely through the group.

The four-year-old female went over to groom her mother. For a few minutes she picked carefully through the long hairs on mother's back. Then she reached for her baby brother and lifted him gently from mother's lap.

She placed the small woolly infant facedown in her own lap and began to groom him. Working with her fingers, she parted the thick hair on his back and examined the exposed skin, searching for small insects and specks of dirt. As she concentrated on her task, she pursed her lips and bent her head close to the baby's body.

At first the baby lay still. Then he began to kick and squirm. Mother reacted instantly. She swept the baby into her arms and nuzzled his shoulder. He reached up and put both arms around her neck.

His sister sat hunched over, staring at the ground. Suddenly she bounced up, galloped across the clearing, threw herself on her side, rolled over, bounded up again, leaped toward a tree and climbed to a lofty branch. She crawled out on the branch and sat with her legs dangling.

Down below she watched two youngsters playing follow-the-leader. One followed the other everywhere — across a fallen log, around a mound, through some low bushes, up a tree and down a hanging vine. Then the leader climbed on top of a tree stump. With arms and legs flying, he tried to keep his playmate from storming the stump.

Across the clearing the sister saw her brother squirming in mother's lap. He was restless, as usual, and wouldn't sit still. As mother held him he reached up and poked at her face with his tiny fingers. She turned her head. Then he arched his back and began to wriggle and twist. Mother lifted him firmly and placed him on the ground.

He stood beside her and raised his hands above his head,

wanting to be picked up again. This time mother ignored him.

Another infant about the same age came bounding over. He reared up on his hind legs and beat his chest with open palms. Then he jumped on the little brother. They tumbled to the ground and wrestled, rolling across the grass with their arms locked around each other.

Finally the brother broke away. He ran past a big male who was sitting quietly on a mound. Whirling around and swinging his arms, the infant swatted the male on the nose. The big gorilla grunted. The baby ran a few feet farther, turned a somersault and ended up on his back. He kicked his legs in the air and waved his arms above his head.

His sister was still watching from her place in the tree. She crawled back along the branch, slid quicky down the tree trunk, ran over to the infant and grabbed for him. He jumped out of the way. She chased him, reaching for his ankles, and finally tackled him. Then she covered him with her body. Twisting and kicking, he struggled to escape.

He squirmed out from under her and sat panting on the ground. A moment later he picked up a stick and began to swing it around. His sister grabbed the other end. They both pulled on the stick until the infant was yanked forward. As they sat face to face, they put their arms around each other and grappled slowly. The baby's mouth was open with the corners pulled back in a wide gorilla smile.

His sister rolled over on her back, and the infant climbed on her belly. She held him with one hand and started to tickle him with the other. He tried to catch her fast-moving hand. When he finally grabbed it, he began to gnaw at her fingers.

As the youngsters played, the adults groomed themselves or napped. One young male was lying on his back. His arms were folded behind his head as he stared up at the sky. Other gorillas had climbed high into the treetops to sun themselves.

Early that afternoon the long siesta ended. The leader of the group walked out to the center of the clearing. He stood with his hands and feet spread, looking straight ahead and grunting as he called the others together.

Gorillas climbed down from the trees and ambled out of the bushes. Infants dashed to their mothers' sides and climbed onto their backs. One newborn infant was still too small and weak to ride piggyback. He clung to his mother's belly as she supported him with one hand. She walked on two feet and the knuckles of the other hand.

The others walked on all fours as they followed their leader into the forest, moving single file along a well-worn path. The four-year-old stayed close behind her mother and brother. She knew this forest trail well, for they had passed this way many times before.

Soon they came to a stream. They walked along the bank until they reached a log that had fallen over the water. Their leader stopped and looked carefully upstream and down. He crossed the log and looked again. Then he stood guard as the rest of them crossed one after another, taking care not to get wet.

Now they spread out for their afternoon feeding. With several others, the four-year-old went to a clump of sprouting bamboo. She pulled a bamboo shoot out of the ground and peeled it like a banana, throwing away the tough outer bark and eating the rest of it.

When she finished the bamboo shoot she wandered away and found an old log lying on the ground. Holding it with both hands, she bit off the rotten bark and licked the creamy fluid that oozed from inside the wood. Another youngster came over. He wiped the log with one finger. Then he licked that finger like a lollipop.

As the two of them bent over the log, they were interrupted by a series of soft, clear hoots. Startled, they looked toward the leader of the group. He was sitting nearby with his head raised and his lips pursed, hooting faster and faster.

Mothers pulled their infants close to them. All the gorillas scrambled to get out of the leader's way. They retreated to a safe distance as he prepared to display his power and strength.

His hoots exploded into a shattering roar. Suddenly he leaped to his feet. He ripped a branch from a tree and hurled it violently into the air. As it left his hand he started to pound on his massive chest, drumming rapidly with cupped palms.

Still pounding on his chest, he ran sideways a few feet. Then he dropped to all fours and dashed wildly across the ground, slapping and tearing at everything in his path. He stopped abruptly, lifted his hand and gave the ground a tremendous whack. After that he settled back quietly. His display was over.

Two other males jumped up to roar and beat their chests. The four-year-old hung back with other youngsters and females, waiting until the males had calmed down. Then all of them hurried to their leader and crowded anxiously around him.

He peered intently into the forest, grunting sharply. Up

ahead, about 100 feet away, two men stood under a tree. Hanging from one man's shoulders were two bright objects — a camera and a pair of binoculars.

The gorillas seemed tense and fearful. For several minutes they huddled by their leader. The four-year-old pressed against her mother's side, watching the men from the corners of her eyes.

The men did not budge. They made no threatening movements. Gradually the gorillas relaxed. As their fear turned to curiosity, they began to crane their necks and stare.

A young male left the group. He slipped into the undergrowth, circled around and then moved cautiously toward the men. He kept low as he crept along, hunching down like a scout. Every few seconds he stood up to look over the tops of bushes. Then he ducked quickly and sat for a moment before creeping on.

He advanced closer and closer. When he was about 30 feet from the men, he reared up, roared loudly and beat his chest. Then he ducked again and peered through the bushes.

Both men were climbing hastily into the tree.

Now the whole group of gorillas began to advance toward the men. They moved slowly and hesitantly, as if daring each other to go closer. The four-year-old tagged along, keeping to the rear as she clutched the hair on her mother's back.

The leader stopped and sat down on a tree stump, with his chin propped on his folded arms. Three gorillas squatted near him, while others crouched behind bushes. The four-year-old finally let go of her mother. With several youngsters she scrambled up into a tree for a better look.

She sat on a branch, peeking through a curtain of hanging vines and biting her lips nervously. Across the way the men crouched in their own tree. When one of them glanced over at her, she ducked her head. As soon as he looked away, she peered out through the vines again.

Down below, the bold young male had moved to within 10 feet of the men's tree. He stopped and stared up at the intruders with his mouth slightly open. Then he slapped the ground with the palm of his hand and looked up again.

The men stayed where they were, staring down at him.

The gorillas were no longer alarmed. One by one they climbed down to the ground and drifted off to munch on greenery. Every so often they glanced up at the men, but mostly they ignored them. The adult males were strong enough to tear a man apart, yet they were not aggressive. By nature they were peaceful and shy. Unless they felt threatened, these gentle giants kept to themselves and bothered no one.

For a while they lingered in the area. Finally the leader of the group grunted softly and moved back into the forest. The others followed. Just before they disappeared among the trees, the leader turned to look back and grunt again. The men still hadn't moved.

For the rest of the day the gorillas ambled leisurely through the forest, stopping here and there to feed. Late that afternoon, while they were resting on the slope of a hill, the four-year-old went over to play with her brother. He climbed on her back and clung tightly to her shoulders as she carried him up and down the hill.

Suddenly their mother came rushing up from behind. She reached out to retrieve the baby, but the sister did not want to give him up. As the baby hung on, the sister galloped around a tree and then crawled into some bushes. Mother chased her. Finally the old female grabbed one of her daughter's ankles.

The sister reached back and swept the baby into her arms. She swung around to face her mother. Whimpering and whining, she clasped the baby to her chest and bobbed up and down until mother pulled the infant away from her.

It was already growing dark. Once again the gorillas gathered around their leader. They sat silently for a while, yawning and scratching as they waited to bed down for the night.

Finally the leader began to break twigs for his nest. That was a signal for the others. They spread out to make their own sleeping nests. Two or three gorillas climbed into trees and built nests among the branches. But most of them nested on the ground.

The four-year-old picked a spot at the foot of a tree. She sat down, reached out for weeds, herbs and other small plants, and bent them in toward her body. Then she stood up and pulled on the mass of greenery, forming a half-circle around her body. She pressed the vegetation down with both hands and broke off twigs that were sticking up. Then she sat down, turned around, and sat again. In a few minutes the nest was ready.

At first she lay on her belly with arms and legs tucked

under. But she was restless. She turned over and pulled her knees up to her chest. Finally she sat up and stared into the darkness.

After a while she climbed out of bed and went over to her mother's nest. Mother was lying on her side with the baby pressed against her. With sleepy eyes she looked up at her daughter. The four-year-old climbed into the nest with them. She snuggled warmly against mother's back.

Just as she was drifting off, she heard a faint thumping sound. She lifted her head to listen. In a nest nearby, one of the gorillas was pounding his chest. He roared briefly in his sleep. Perhaps he was dreaming.

For Further Reading

Francis Downes Ommanney, *Frogs, Toads, and Newts* (McGraw-Hill, N.Y., 1975)

William White, Jr., *A Frog Is Born* (Sterling, N.Y., 1972)

Herbert S. Zim, *Frogs and Toads* (William Morrow, N.Y., 1950)

Barbara Brenner, *A Snake-Lover's Diary* (Young Scott, Reading, Mass., 1970)

Bessie Hecht, *All About Snakes* (Random House, N.Y., 1956)

Hilda Simon, *Snakes: Facts and Folklore* (Viking, N.Y., 1973)

Johanna Johnston, *The Eagle in Fact and Fiction* (Crown, N.Y., 1966)

Sigmund A. Lavine, *Wonders of the Eagle World* (Dodd, Mead, N.Y., 1974)

Robert Whitehead, *The First Book of Eagles* (Franklin Watts, N.Y., 1968)

Irmengarde Eberle, *Beavers Live Here* (Doubleday, N.Y., 1972)

Walter Edmonds, *Beaver Valley* (Little Brown, Boston, 1971)

Glen Rounds, *Beaver Business* (Prentice Hall, Englewood Cliffs, N.J., 1960)

Pat Cherr, *The Lion in Fact and Fiction* (Harlin Quist, N.Y., 1966)

Herbert S. Zim, *The Big Cats* (William Morrow, N.Y., 1955)

David Stephen, *Cats* (Putnam, N.Y., 1975)

Robert Gray, *The Great Apes* (Norton, N.Y., 1969)

Helen Kay, *Apes* (Macmillan, N.Y., 1970)

Alice Schick, *Kongo and Kumba: Two Gorillas* (Dial, N.Y., 1974)

AUTHOR'S NOTE: The accounts given here of lion and gorilla social life are based on George Schaller's landmark studies, *The Serengeti Lion* (University of Chicago Press, 1972) and *The Mountain Gorilla* (University of Chicago Press, 1963).

591.3   **Freedman, Russell**
F   Growing up wild

                                    X49349